GLACIAL EPOCHS AND WARM

POLAR CLIMATES

(1879)

BY

ALFRED RUSSEL WALLACE

British Library Cataloguing-in-Publication Data
A catalogue record for this book is available from the
British Library

Alfred Russel Wallace

Alfred Russel Wallace was born on 8th January 1823 in the village of Llanbadoc, in Monmouthshire, Wales.

At the age of five, Wallace's family moved to Hertford where he later enrolled at Hertford Grammar School. He was educated there until financial difficulties forced his family to withdraw him in 1836. He then boarded with his older brother John before becoming an apprentice to his eldest brother, William, a surveyor. He worked for William for six years until the business declined due to difficult economic conditions.

After a brief period of unemployment, he was hired as a master at the Collegiate School in Leicester to teach drawing, map-making, and surveying. During this time he met the entomologist Henry Bates who inspired Wallace to begin collecting insects. He and bates continued exchanging letters after Wallace left teaching to pursue his surveying career. They corresponded on prominent works of the time such as Charles Darwin's *The Voyage of the Beagle* (1839) and Robert Chamber's *Vestiges of the Natural History of Creation* (1844).

Wallace was inspired by the travelling naturalists of the day and decided to begin his exploration career collecting specimens in the Amazon rainforest. He explored the Rio Negra for four years, making notes on the peoples and languages he encountered as well as the geography, flora, and fauna. On his return voyage his ship, Helen, caught fire and he and the crew were stranded for ten days before being picked up by the Jordeson, a brig travelling from Cuba to London. All of his specimens aboard Helen had been lost.

After a brief stay in England he embarked on a journey to the Malay Archipelago (now Singapore, Malaysia, and Indonesia). During this eight year period he collected more than 126,000 specimens, several thousand of which represented new species to science. While travelling, Wallace refined his thoughts about evolution and in 1858 he outlined his theory of natural selection in an article he sent to Charles Darwin. This was published in the same year along with Darwin's own theory. Wallace eventually published an account of his travels *The Malay Archipelago* in 1869, and it became one of the most popular books of scientific exploration in the 19th century.

Upon his return to England, in 1862, Wallace became a staunch defender of Darwin's landmark work *On the Origin of Species* (1859). He wrote responses to those critical of the theory of natural selection, including 'Remarks on the Rev. S. Haughton's Paper on the Bee's Cell, And on the Origin of Species' (1863) and 'Creation by Law' (1867). The former of these was particularly pleasing to Darwin. Wallace also published important papers such as 'The Origin of Human Races and the Antiquity of Man Deduced from the Theory of 'Natural Selection" (1864) and books, including the much cited *Darwinism* (1889).

Wallace made a huge contribution to the natural sciences and he will continue to be remembered as one of the key figures in the development of evolutionary theory.

Wallace died on 7th November 1913 at the age of 90. He is buried in a small cemetery at Broadstone, Dorset, England.

GLACIAL EPOCHS AND WARM POLAR CLIMATES

When the late Professor Agassiz, in the year 1840, discovered clear traces of glacial action in the valleys of Scotland, his announcement was received with incredulity, even by geologists. The whole body of evidence afforded by fossil remains was supposed to prove that our climate had formerly been warmer than it is now, and that the farther we went back in geological time the more tropical were the forms of life which inhabited Europe. This was seen to be entirely consistent with the theory of a cooling earth, to which the high temperatures of the earlier periods were then almost universally attributed; and, as the idea of a much colder climate in former times did not harmonize with this theory, the glacialists were for some time looked upon as a set of wild enthusiasts, whose facts were not worth examining, and whose theories might be ridiculed or despised. Soon, however, the tide began to turn. Eminent geologists, after visiting the alpine valleys where glaciers are still at work, were struck with the identity of the phenomena evidently produced by them with those to be still seen in all our mountain regions; the great importance of this identity was acknowledged; and thenceforth a body of enthusiastic and industrious workers

7

arose, who have made this phase of the ancient history of our earth their special study. With admirable patience they have tracked the ice-marks far and wide over the northern hemisphere, and have so skilfully interpreted the mysterious record, that we are now able to read with confidence the great outlines of this most marvellous chapter in the past history of our earth--the Glacial Period. More recently, the subject has been greatly extended by the growing belief that similar cold epochs have occurred during tertiary, secondary, and even palæozoic times--a belief founded on numerous facts which had previously been overlooked or misinterpreted; while the proofs of a corresponding succession of warm periods within the Arctic regions have accumulated so rapidly, that no doubt now remains as to this still more startling phase of climatic change.

To account for this wonderful series of phenomena the old theory of a cooling earth is evidently inadequate, and has long been given up by geologists, while it has been shown by physicists to be altogether untenable. Another theory has however taken its place, which, though beset with many difficulties, does really account for almost the whole of the series of facts above referred to, and is therefore adopted with more or less reserve by most students of the subject. But the facts are so varied and widespread, their interpretation is sometimes so obscure, and the theory which explains them is often so difficult in its application, that the whole question appears to most persons to be a hopeless puzzle, while many even doubt the fundamental fact of there ever having been glacial epochs of the intensity and wide extent usually claimed for them. Within the last few years, however, much light has been thrown on the subject by the various Arctic expeditions and by discussions on oceanic circulation, and the time has probably now come when a condensed and intelligible account, both of the facts and of the best mode of accounting for the facts, will be acceptable to a wide circle of intelligent readers. Such an account will here be attempted.

The Glacial Epoch.--The phenomena that prove the comparatively recent occurrence of a glacial epoch in the temperate regions of the northern hemisphere are exceedingly varied and extend over very wide areas. It will be well therefore to state, first, what these facts are as exhibited in our own country, referring afterwards to corresponding phenomena in other parts of the world.

Perhaps the most striking and conclusive among these facts are the grooved, scratched, or striated rocks. These occur abundantly in Scotland, the Lake country, and North Wales, and no rational explanation of them has yet been given, except that they were formed by glaciers. In many valleys, as for instance that of Llanberis in North Wales, hundreds of examples may be seen of deep grooves several inches wide, smaller furrows, and striæ of extreme fineness, whenever the rock is of sufficiently close texture and sufficiently hard to receive and retain such marks. These grooves or scratches often extend for many yards in length; they are found in the bottom of the valley as well as high up on its sides, and they are all (almost without exception) in one direction--that of the valley itself--even though the particular surface they are upon slopes in another direction. Where the native covering of turf is freshly removed, the grooves and striæ are often seen in great perfection, and there is reason to believe that such markings cover, or have once covered, a large part of the rocky surfaces of the countries where they are now found. Accompanying these striæ we find another phenomenon, hardly less curious,--the rounding off or planing down of the hardest rocks to a smooth undulating surface. Hard crystalline schists, with the strata nearly vertical, and which one would expect to find exposing jagged edges, are often ground off to a smooth, but never to a perfectly flat surface. These smoothly-rounded surfaces are found not only on single rocks, but over whole valleys and mountain-sides, and form what are termed *roches moutonnées,* from their having the appearance at a distance of sheep lying down.

Now these two phenomena are actually produced elsewhere by existing glaciers, while there is no other known or even conceivable agency that could have produced them. Whenever the Swiss glaciers retreat a little, as they sometimes do, the rocks they have moved over are found to be rounded, grooved, and striated just like those we now see in Wales and

Scotland. The two sets of phenomena are so exactly identical, that no one who has ever compared them can doubt that they are due to the same causes. But we have further and even more convincing evidence. Glaciers produce many other effects besides these two; and whatever effects they now produce in Switzerland, in Norway, or in Greenland, we find examples of similar effects having been produced in our own country. The most striking of these are moraines and travelled blocks, which must be briefly described.

Almost every existing glacier carries down with it great masses of rock, stones, and earth, which fall on its surface from the precipices and mountain slopes which hem it in. As the glacier slowly moves downward, this debris forms long lines on each side, or on the centre whenever two glacier-streams unite, and is finally deposited at its termination in a huge mound called the terminal moraine. The decrease of a glacier may often be traced by successive old moraines across the valley up which it has retreated. When once seen and examined in connection with an existing glacier, these moraines can always be distinguished almost at a glance. Their position is most remarkable, having no apparent relation to the valley or the surrounding slopes. They look like huge earthworks formed for purposes of defence, rather than works of nature; and their composition is equally peculiar, consisting of a mixture of earth and rocks of all sizes, without any arrangement whatever, the rocks often being huge angular masses just as they had split off the surrounding precipices. Some of these rocky masses often rest on the very top of the moraine, in positions where no other natural power but that of ice could have placed them. Exactly similar mounds are found in the valleys of Wales and Scotland, and always in districts where other evidences of ice-action can be abundantly traced, and where the course of the glacier which produced them can be readily understood.

The phenomenon of travelled or perched blocks is also a common one in all glacier-countries, marking out very clearly the former extension of the ice. The glacier which fills a lateral valley will often cross over the main valley, and abut against the opposite slope, and will deposit there some portion of its terminal moraine. But in these cases the end of the glacier will spread out laterally and the moraine matter will be distributed over a large surface, so that the only well-marked feature left by it will be some of the larger masses of rock that may have been brought down. The same thing will occur when a glacier surrounds an isolated knoll or overrides a projecting spur, on both of which large blocks brought from a distance may be left stranded. Such blocks are found abundantly in many districts of our own country where other marks of glaciation exist, and they often consist of a rock different from that on which they rest, or from any in the immediate vicinity. They can, however, almost always be traced to their source in one of the higher valleys from which the glacier descended; and this remoteness of origin, combined with their great size, their angular forms, and their singular positions, often perched on the crest of a ridge, on a steep slope, or on the summit of a knoll, altogether preclude any other known mode of transport but by glaciers or floating ice.

Some of the most remarkable examples of these travelled blocks are to be seen on the southern slopes of the Jura mountains. They consist of enormous angular masses of granite, gneiss, and other crystalline rocks, quite foreign to the Jura range which consists entirely of Jurassic limestones and tertiary deposits, but exactly agreeing with those of the main Alpine range, more than fifty miles away across the great central valley of Switzerland. These blocks have been proved by Swiss geologists to have been brought by the ancient glacier of the Rhone, which was fed by the snows of the whole Alpine range, from Mont Blanc to Monte Rosa and the head of the Rhone valley, a distance of nearly 120 miles, together with the southern slopes of the Bernese Alps--a district which comprises all the most extensive snowfields and glaciers in Switzerland. The whole area between these two ranges, which may be roughly described as about 100 miles long by 30 wide and bounded by mountains from 7000 to 10,000 feet high, must have been literally filled with ice; and this enormous glacier discharged itself at the mouth of the upper Rhone-valley, between the Dent de Morcles and the Dent de Midi, in an ice-stream of such enormous thickness as to fill up the bed of the Lake of Geneva, and, though spreading widely over the valley, retaining a thickness of more than 2000 feet when it reached the Jura.

The blocks brought by it are found scattered over the slopes of the Jura for a distance of about 100 miles, and it is a very interesting fact, that they reach the greatest height in a direct line with the course of the glacier, while on both sides of this point they are found lower and lower. One of the largest of the blocks is forty feet in diameter, and many others are of enormous size. They vary considerably in the material of which they are composed; but they have each been traced to their source, which has always been found in a part of the Alpine range corresponding to their actual position on the theory that they were brought, as described, by the great Rhone glacier. Thus, all the blocks found to the east of a central point near Neufchatel can be traced to the eastern side of the Rhone valley, while those found towards Geneva have all come from the west side. It is evident that, had these blocks been carried by floating ice during a period of submergence (as was at first supposed), their distribution must have been different. There could have been no accurate separation of those derived from the opposite sides of the same valley, while all would have been stranded at the same elevation, or in parallel lines indicating the different heights at which the water stood at different epochs. These considerations are so weighty, that they compelled Sir Charles Lyell to withdraw the explanation he first gave--of the carriage of the blocks by floating ice during a period of submergence--as quite untenable, and to accept, as the only explanation which covered the facts, the enormous glacier above described.[1] Similar phenomena, though nowhere on so grand a scale, are found in the vicinity of all the mountain ranges of Central Europe, and they undoubtedly demonstrate the fact of a recent change of climate, from a time when all our higher mountains were covered with perpetual snow, and the adjacent valleys were filled with glaciers at least as extensive as those now found in Switzerland.

But this conclusion, marvellous as it is, by no means affords us an adequate conception of the condition of our islands during the whole duration of the glacial epoch; for there are other phenomena, best developed in Scotland, which show that what we have hitherto described was but its concluding phase, and that during its maximum development the mountainous parts of our islands must have been reduced to a condition only to be now paralleled in Greenland and the Antarctic regions. As few persons besides professed geologists are acquainted with the weight of the evidence in support of this statement, we will here briefly summarize it, referring our readers to Mr. Geikie's volume for fuller details.

Over almost all the lowlands, and in many of the highland

valleys of Scotland, there are immense superficial deposits of clay, sand, gravel, or drift, which can be traced more or less certainly to glacial action. Some of them are moraines, others are lacustrine deposits in temporary lakes formed by moraines or glaciers, while others have been formed or deposited under water during periods of submergence. But below them all, and often lying directly on the solid rock, there are extensive layers of a very tough clayey deposit called 'till.' It is very fine in texture, very tenacious, and often of a stone-like hardness; and it is always full of rocks and stones, which are of rude sub-angular forms, rubbed smooth and partially rounded, and almost always covered with scratches or deep striæ, often crossing each other in various directions. Sometimes the stones are so numerous, that there seems to be only just enough clay to unite them into a solid mass; and they are of all sizes, from mere grit up to rocks many feet in diameter. The 'till' is found chiefly in low-lying districts, where it covers extensive areas, sometimes to the depth of a hundred feet; while in the highlands it occurs in much smaller patches, but in some of the broader valleys it forms terraces which have been cut through by the streams. Occasionally it is found as high as two thousand feet above the sea in hollows on hillsides, or in other situations where it seems to have been protected from denudation.

The 'till' is totally unstratified, the stones it contains being found all mixed together and evidently unsorted by water, while the rock surfaces on which it rests are invariably worn smooth, and greatly grooved and striated when the rock is hard, while when it is soft or jointed it frequently has a greatly broken surface, as if it had been subjected to enormous pressure. The colour and texture of the 'till' and the nature of the stones it contains all correspond with the character of the rock of the district, so that it is clearly a local formation. It is often found underneath moraines, drift, and other glacial deposits, but never overlies them, so that it was certainly formed at an earlier date.

Throughout Scotland, where 'till' is found, the glacial striæ, perched blocks, *roches moutonnées,* and other marks of glacial action, occur very high up the mountains, to at least 3000 and often even to 3500 feet above the sea; while all lower hills and mountains are rounded and grooved up to their very summits, and these grooves always radiate outwards from the highest peaks and ridges towards the valleys or the sea.

Now these various phenomena taken together are only explicable on the supposition that the whole of Scotland was once buried under a vast ice-sheet, above which only the highest mountains reared their summits. There is absolutely no escape from this conclusion, for the facts which lead to it are not local, found only in one spot or one valley, but general throughout the entire length and breadth of Scotland; and these facts correspond so wonderfully in every detail to this conclusion, and this only, as to amount to absolute demonstration. The weight of this vast ice-sheet, at least three thousand feet in maximum thickness, and continually moving seaward with a slow grinding motion like that of all existing glaciers, must have worn down the whole surface of the country, especially all the prominences, leaving the grooves and striæ we still see, marking the direction of its motion. All the loose stones and rocks which lay on the surface would be pressed into the ice, the harder ones would serve as scratching and grinding tools, and would thus themselves become worn, scratched, and striated as we find them, while all the softer masses would be ground up into impalpable mud along with the material worn off the rock-surfaces of the country.

The peculiar characters of the 'till,' its fineness and tenacity, correspond closely with the fine matter which now issues from under all glaciers, making the streams milky white, yellow, or brown, according to the nature of the rock. The sediment from such water is a fine unctuous ooze, only needing pressure to form it into a tenacious clay; and when 'till' is exposed to the action of water it dissolves into a similar soft, sticky, unctuous mud. The present glaciers of the Alps, being confined to valleys which carry off a large quantity of drainage water, lose this mud perhaps as rapidly as it is formed; but when the ice covered almost the whole country, there would be comparatively little drainage water, and the mud and stones would collect in vast compact masses in sheltered hollows, and especially in the lower flat valleys, which would necessarily be ground into hollows or basins where the pressure of the ice was greatest. As the ice retreated, the areas of greatest pressure would retreat also, and the hollows would be left full of the stones and glacier-mud which was continually being formed. At a later period it was greatly denuded by rain and rivers, but, as we have seen, large quantities remain to this day, to tell the wonderful story of its formation. It was at this time that the glaciers of Wales and of Ireland acquired their greatest extension, and there is clear evidence that the ice on the west of Scotland extended far out to sea, overspreading all the islands, and connecting itself in one unbroken mass with the almost equally extensive ice-sheet that covered Ireland.

It is evident that the great change of climate, which produced such marvellous effects in the British Isles, could not have been confined to them; and accordingly we find that there are strikingly similar proofs that Scandinavia and much of Northern Europe have also been covered with a continuous ice-sheet. We have already seen that huge glaciers almost buried the Alps, carrying granitic blocks to the Jura and depositing them on its flanks to a height of 3450 feet above the sea-level; while to the south, in the plains of Italy, the terminal moraines left by the retreating glaciers have formed considerable hills, those of Ivrea, the work of the great glacier of the Val d'Aosta, being fifteen miles across, and from 700 to 1500 feet high.

In North America the marks of glaciation are even more extensive and perhaps more remarkable than in Europe, stretching over the whole eastern part of Canada, and at least as far as the 40th parallel of latitude, south of the great lakes. There is, over considerable areas, a deposit like the 'till' of Scotland, produced by the grinding of the great ice-sheet when it was at its maximum thickness. In the eastern part of Canada and the United States, the ice appears to have risen to its greatest height over the northern watershed of the St. Lawrence near its mouth, and to have extended across to the White Mountains of New England, filling up the Gulf of St. Lawrence, overspreading Nova Scotia and Long Island, and terminating in an ice-cliff in the Atlantic. It is believed by Mr. J. D. Dana to have had a thickness of over 5000 feet in New England, and to have reached a height of 13,000 feet over the northern watershed of the St. Lawrence. At a later period the local glaciers left moraine-matter, travelled-blocks, and striated rocks, as in Europe. There are also in North America, as well as in Britain and Scandinavia, proofs of the submersion of the land beneath the sea, to a depth of more than a thousand feet, in the latter part of the glacial epoch; but this is a subject we need not here enter upon, as our special object is to show the reality and magnitude of that wonderful and comparatively recent change of climate termed the glacial epoch.

Many educated persons, and even men of science who have not given much attention to the subject, look upon the whole of this enquiry as an elaborate but highly improbable theory, made to fit certain phenomena which are otherwise difficult to explain; and they would not be much surprised if they were some day told that it was all a delusion, and that somebody had explained the whole thing in a more simple way. The outline now given of the wide basis of facts on which the theory rests, will, it is hoped, prevent any of our readers from being imposed upon by such statements or disturbed by any sceptical doubts. There is perhaps no great conclusion in any science, which rests upon a surer foundation than this does; and if we are ever to be guided by our reason in deducing the unknown from the known, or the past from the present, we cannot refuse our assent to the reality of the glacial epoch in all its more important features. Just as surely as the extinct volcanoes of Auvergne or of Victoria demonstrate the former existence of active volcanoes and flowing lava-streams, so surely do the striated rocks, the perched blocks, the moraines, and the 'till,' demonstrate the former existence of the glaciers, by which alone they could have been produced.

Before quitting this part of our subject, we must notice some curious facts, which seem to show that there were recurring periods of warmth during the glacial epoch itself, as they have a very important bearing on the theory by which changes of climate in general seem to be best explained.

The 'till,' as we have seen, could only have been formed when the country was buried under an ice-sheet of enormous thickness, and when in the parts so buried there could have been neither animal nor vegetable life. But in several places in Scotland, layers of sand and gravel with beds of peaty matter have been found intercalated between layers of 'till.' Sometimes these intercalated beds are very thin, but in other cases they are twenty or thirty feet thick, and contain the remains of the extinct ox, the Irish elk, the horse, reindeer, and mammoth. Here then we have evidence of two distinct periods of intense cold when the country was buried in ice, and of an intervening mild period sufficiently prolonged for the country to become covered with vegetation and stocked with large animals. In some districts borings have proved the existence of no less than four distinct formations of 'till,' separated from each other by beds of sand from two to twenty feet in thickness.[2] In North America similar beds occur, intercalated between true glacial deposits, and containing remains of the elephant, mastodon, and great extinct beaver. In Switzerland a similar interglacial bed contains peat, and remains of pines, oaks, birches, and larch, with bones of the elephant, rhinoceros, stag, and cave-bear, and also abundance of insects.[3] There seems, therefore, to be ample proof that the glacial epoch did not consist of one continuous change from a temperate to a cold and arctic climate, which, having reached a maximum, then passed slowly and completely away, but that there were certainly two, and probably many, distinct alternations of arctic and temperate climates.

It is true that the evidence of such alternations is scanty, but a little consideration will show that we could not expect to find more complete evidence, because each succeeding ice-sheet would necessarily grind down or otherwise destroy most of the superficial deposits left by its predecessors, while the torrents that must have been produced by the melting of the ice would wash away most of the fragments which had escaped. It is therefore fortunate that we find *any* portions of interglacial deposits containing vegetable and animal remains; and, as we might expect, these seem to have been formed when each succeeding phase of the cold period was less severe than those which preceded it, in other words, when the glacial epoch was passing away. If there had been similar intercalated warm periods while it was coming on, it is hardly possible that any record of them could have been preserved, because each succeeding extension of the ice would be greater than that which preceded it, and would certainly destroy all traces of animal or vegetable remains in superficial deposits.

Now it is a very remarkable fact, that the only theory which affords a distinct explanation of the changes which brought about the glacial period, leads to the conclusion that such alternations of warm and cold climates must have occurred during its continuance; and as the same theory also explains, more or less completely, the other changes of climate of which we have geological evidence, it will be most convenient to our readers at once to give an outline of that theory, and to show as clearly as possible by its aid how the glacial epoch was brought about. In doing so we shall have the opportunity of clearing up some of the misconceptions with which the subject has been encumbered; and we shall then proceed to give an account of the proofs of warm arctic climates, and to show how far the same theory is able to explain them.

Astronomical causes of periodical changes of Climate.--The elliptical orbit in which the earth moves round the sun has an eccentricity of about one and a half millions of miles, which causes us to be sometimes three millions of miles nearer the sun than at others. Strange as it may seem, we in the northern hemisphere are nearest to the sun in winter, and furthest off in summer, while in the southern hemisphere it is the reverse; and this peculiarity must have some effect in making our northern winters less severe than those of the Antarctic lands. But the earth moves more rapidly in that part of its orbit which is nearest the sun, so that our winter is not only milder, but several days shorter, than that of the southern hemisphere. There can be no doubt that if this state of things were reversed, our winters occurring when we were furthest from the sun, or in *aphelion*, and our summers when we were nearest the sun, or in *perihelion*, we should experience a decided difference in climate, our winters being more severe and longer, and our summers hotter but shorter. Now there is a combination of astronomical causes (the precession of the equinoxes and the revolution of the apsides) which actually brings this change about every 10,500 years, so that after this interval the condition of the two hemispheres is reversed as regards nearness to the sun in summer and comparative duration of summer and winter, and this change has been going on throughout all geological periods. As we shall continually have to refer to these changes, we shall speak of them as the 'opposite phases of precession,' when winter in the northern hemisphere occurs at *aphelion* or *perihelion*, that is when the earth is farthest from or nearest to the sun.

But the amount of eccentricity itself varies very largely, though very slowly, and it is now nearly at a minimum. It also varies very irregularly, but its amount has been calculated by Mr. Croll and others for 3,000,000 years back, during which time it has much more frequently been of greater than of less amount than it is now. Going backward from the present time, we find that for about 25,000 years it was slightly greater, and for 30,000 years more rather less, than it is now, so that for a period of nearly 60,000 years back we can impute no important change of climate to this cause.

But it then increases rapidly till about 72,000 years ago, when it was double, while 100,000 years ago it was more than two-and-a-half times, its present amount. Going back still further, it diminishes for 50,000 years to double, and then again increases for 60,000 years till it becomes nearly three-and-a-half times what it is now. It then diminishes rapidly, and after one more considerable rise, sinks to almost exactly its present amount 400,000 years ago. We thus find that for a period of about 140,000 years, beginning 70,000 years back, the eccentricity was always more than double its present amount, with two maxima, of more than two-and-a-half and nearly three-and-a-half times that quantity. It is to this long period of high eccentricity that geologists are pretty generally agreed to refer the glacial epoch, and to this we shall accordingly for the present confine our remarks.

Let us then endeavor to ascertain what would have been the climatic condition of our country at the period of greatest eccentricity 210,000 years ago. The mean distance of the sun may be taken at 92,500,000 of miles, and the eccentricity at the period we are referring to was, in round numbers, five millions of miles. The sun's distance would therefore be 97,500,000 miles when in *aphelion*, and 87,500,000 when in *perihelion*--a difference of ten millions of miles. The question is, what difference would this make in our climate, when our winters occurred in *aphelion* or at a time when we were furthest from the sun? And in estimating this we must remember that the quantity of heat received from the sun is in inverse proportion to the square of the distance, so that instead of being about one-tenth less, as the simple proportion of the above numbers would give, the difference will be very nearly one-fifth--that is, the earth would be receiving only four-fifths of the heat in *aphelion* that it received in *perihelion*.

In order to understand what effect this would have on climate, we must first obtain a clear idea of the power of the sun in raising the earth's surface-temperature; and on this subject there is a very general misconception, owing to the unscientific manner in which all thermometers are graduated and figured, so that we reckon degrees of temperature either from the freezing-point of water or from a point thirty-two degrees below it. The true zero from which we ought to measure terrestrial temperatures is, however, about 271° F. below the freezing point, or -239° F., this being the temperature to which the surface of the earth would sink if the heat of the sun were entirely withdrawn from it. When therefore we speak of the mean temperature of the equator being 80° F., while that of England is about 50° F., these numbers convey no real information, either as to the total amount of sun-heat received or as to its proportions in the two latitudes. But if we add to both of these numbers 239° F., we obtain 319° F. for the absolute temperature of the equator, and 280° *sic* F. for that of England, and these numbers are truly proportional to the amount of heat in each place due, directly or indirectly, to the solar influence. We may here mention that it is the universal opinion of modern physicists, that the internal heat of the earth escapes outward so slowly as to contribute no sensible portion of that which we perceive as climatic temperature, and that it may therefore be neglected in all calculations as to comparatively recent changes of climate. Two facts may be adduced, which, if they do not prove, are at all events quite in accordance with this view, while they are opposed to the view that the internal heat makes itself felt in any important degree. The one is, that in the very deepest tropical oceans the bottom temperature is close to the freezing-point; the other, that in many parts of Siberia, after passing below the surface-soil, there are found seventy and in some places several hundred feet of permanently frozen ground.

The mean January temperature of England may be taken at 39° F., which is equivalent to 278° F. of absolute temperature; and if we calculate what would be the mean temperature of the same month, when the sun was distant 97,500,000, instead of 91,000,000 of miles as it is now, we find it comes out 242° F., which is equivalent to 3° F. of our thermometers, or 29° of frost. Calculated in the same way, the summer would be excessively hot, the July mean temperature being 125° F. The winter, however, would be 26 days longer than the summer, and the total quantity of sun-heat received by us during the year would be exactly the same as it is now. How then, it may be asked, could this produce a glacial period? Would not the snow that fell in the winter be all melted by the intensely hot summer? The answer to these questions involves considerations of great interest, which must be discussed in some detail; but we may first point out, that the figures above given are not to be taken as implying that these temperatures would actually occur, but rather as quantities of heating power, which might be distributed over wider areas or longer times, so as to produce results differing considerably from the maxima and minima indicated. Our actual climate does not depend so much on the amount of sun-heat we ourselves receive, as on the way in which the heat of adjacent regions is distributed. Owing to our insular position and the influence of the Gulf-stream, we have a much warmer and more uniform climate than is due to our latitude, yet the cold arctic winds often lower our temperature to the freezing-point, at times when we are receiving an amount of sun-heat which would give us genial warmth if its effect were not thus counteracted.

At the time we are speaking of, the winter of the whole northern hemisphere would be as much colder on the average as our own would be, that is, about 36° F. colder than now. The tropics and parts of the southern hemisphere would also be colder, so that it is almost certain that our climate in winter would have been quite as bad if not worse than these figures represent, and would probably have been something like that of Yakoutsk in Northern Siberia, the coldest place in the world outside the Arctic circle. It is hardly possible, therefore, to exaggerate the degree of winter cold of Britain and of all the north temperate zone at that period; and the winter would not only be thus intensely cold, but all its effects would be exaggerated by its being very long. Then would follow a summer of great heat, with a sun probably as fierce as that of India or Australia, and which, though shorter than our present summers by nearly a month, would still be long as compared with that of the Arctic regions. And during this summer, we must remember, all the world would be warmer; evaporation would be greater than now, and the surface waters of the ocean, being kept at a high temperature, would give us warmer and moister breezes than we now possess. Again, therefore, we have to ask, could the winter do anything that the summer could not undo? In order to be in a position to answer this question, we must consider how the heat received from the sun is disposed of, and the influence of water and air, snow and ice, in distributing or storing up heat or cold.

The great aerial ocean which surrounds the earth has the wonderful property of allowing the heat-rays from the sun to pass through it without being warmed by them. When the earth is heated, however, the air gets warm by contact with it, and the vapour and carbonic acid in the air are also impervious to the radiant heat given out by the earth, and are therefore warmed by radiation. But the air thus warmed is in continual motion, owing to change of density. It is lifted up and pushed aside by cooler and heavier air; and thus heat can never *accumulate* in the atmosphere, or warm it beyond a very moderate degree, so that the long-continued sun-heat of the tropics is in great part carried away to give warmth to colder regions. Water also is mobile; and though it receives and stores up a great deal of heat, it is for ever dispersing it over the earth. The rain, which brings down a certain portion of heat from the atmosphere or absorbs it from the earth on which it falls, flows away in streams to the ocean; while the ocean itself, constantly impelled by the winds, forms great currents, which carry off the surplus heated water of the tropics to the temperate and even to the Polar regions. An immense quantity of sun-heat is also used up in evaporating water, and the vapour is conveyed by the aerial currents to distant countries, where, on being condensed into rain, it gives up much of this heat to the earth and atmosphere. The power of water in carrying away heat is well seen in the abnormally high temperature of arid deserts, while the still more powerful influence of air can be best understood by considering how rapidly a few hours of our northern sunshine will heat a tightly-closed glasshouse far above the temperature produced even by the vertical sun of the equator where the air is free to circulate. We can quite understand, then, that, were not a very large proportion of the sun's heat carried away from the tropics by air and water, those parts of the earth would be uninhabitable furnaces, as would indeed *any* part of the earth where the sun shone brightly throughout a summer's day.

But though sun-heat cannot be stored up beyond a very moderate amount, yet cold can be stored up to an almost unlimited extent, owing to the peculiar property of water in becoming solid at a very moderately low temperature; and as this is a point of the highest importance in the enquiry we are now pursuing, it must be carefully considered.

Let us then examine the very different effects produced by water falling as rain, or as snow, which may yet not differ from each other more than two or three degrees in temperature. The rain, however much of it may fall, runs off rapidly into rivers, and soon reaches the ocean. If cold, it cools the air and the earth temporarily, but can produce no permanent effect, because a few hours or days of sunshine restore to the air and the surface soil all the heat they had lost. But if snow falls for a long time, it remains where it fell, becomes compacted into a mass, and keeps the earth below it and the air above it at or very near the freezing-point till all is melted. If the quantity is great, it may take days, or weeks, or months, to melt it all; and if more falls during winter than can be melted during the succeeding summer, then the snow will be perpetual, as we find it above a certain height on all great mountains, and in some parts of the Arctic and Antarctic regions, and then no amount of sun-heat warms the air much above the freezing-point. In N. latitude 80°, Captain Scoresby had the pitch on one side of his ship melted by the heat of the sun, while salt water was freezing on the other side, owing to the coldness of the air.

It is very important to remember that this storing up of cold depends entirely on the amount of snowfall and the want of sufficient heat to melt it, not at all on the actual cold of the winter or the average cold of the year. A place may be intensely cold in winter and may even have a very short summer, yet if so little snow falls that it is quickly melted, there is nothing to prevent that summer being warm and the earth producing a rich and luxuriant vegetation. As examples of this, we have great forests in the north of Asia and America, where the winters are colder and the summers shorter than in southern Greenland, or in Heard Island, both of which are almost wholly covered with perpetual snow and ice. At the Jardin, on the Mont Blanc range, surrounded by perpetual snow, the lowest winter temperature is -6° F., while at Upernivik in Greenland it is -49° F., yet in summer the thermometer often rises to 50° F. or even to 58° F., and the country is covered with a luxuriant herbaceous and shrubby vegetation. Even in the very highest latitudes reached by our last Arctic expedition, there is very little perpetual snow or ice; for Captain Nares tells us, that north of Hayes Sound, in latitude 79° N., the mountains were remarkably free of ice-cap, while extensive tracts of lands were free from snow during the summer.[4] And the reason of this is evidently the scanty snowfall, which rendered it sometimes difficult to obtain enough to form protecting banks round the ships; and this was north of 80° latitude, where the sun was absent for 142 days.

It is a very remarkable and most suggestive fact, that nowhere in the world at the present time are there any extensive lowlands covered with perpetual snow. The Tundras of Siberia and the barren grounds of North America are all clad with some kind of summer vegetation; and it is only where there are lofty mountains or high plateaux, as in Greenland, Spitzbergen, and Grinnell Land, that glaciers, accompanied by perpetual snow, descend to the sea-level. In the Antarctic regions there are extensive highlands, and these are everywhere exposed to the influence of the moist sea air; and it is here accordingly that we find what appears to be a veritable belt of permanent ice covering the whole circumference of the Antarctic continent, and forming a girdle of ice-cliffs which almost everywhere descend into the sea. Such Antarctic islands as South Georgia, South Shetland, and Heard Island, are often said to have perpetual snow at the sea-level; but they are all mountainous and send glaciers down into the sea, and as they receive moist air on every side, the precipitation, which almost all takes the form of snow even in summer, is of course unusually large.

Lieutenant Payer of the Austrian Arctic expedition has furnished us with a clue to these phenomena, by his observation of the rate of evaporation of ice during the Arctic summer. He found that no less than four feet thick of ice was thus evaporated. Now to replace this, a quantity of snow equivalent to about forty-five inches of rain must fall; and as the precipitation rarely reaches this amount outside the tropics, except in islands and in mountainous districts, and it is generally less than half as much over the greater part of the northern continents, we can quite understand why perpetual snow should be so rare a phenomenon on lowlands. Mountains, however, acting as condensers, are exposed to great precipitation, and when covered with snow their condensing power becomes greater, and this tends to increase the accumulation of snow; while the rarefied air has less capacity for heat, and is thus unable to melt or evaporate the snow as it does on the lowlands. We can thus understand how it is that only where there are extensive highlands, to act both as condensers of moisture and accumulators of snow, do we find anything approaching the condition of things which appears to have prevailed in our country during the glacial epoch.

These considerations seem to render Mr. Croll's theory of polar ice-caps, constantly increasing in thickness to the pole itself, quite untenable. He believes that the Antarctic continent is now covered by such an ice-cap, the thickness of which at the South Pole he calculates at twelve miles, giving a slope at an angle of only half a degree,--and less than this, he says, will not produce motion. To the objection, that the outer band of ice will condense all the vapour, he replies that however small the annual amount of snowfall may be, if more falls than is melted, the ice must continue to accumulate year by year, till its thickness in the centre of the continent be sufficiently great to produce motion.[5] But all the evidence we have of the rapid diminution of snowfall and rainfall inland, and especially when the coast consists of lofty mountains, as in the Antarctic regions, combined with the enormous evaporating power of the four months' polar sun, render such continuous increase as he supposes quite impossible. He has himself told us that during summer, from May 10th to August 3rd, a period of eighty-five days, the quantity of heat received from the sun in consequence of his remaining above the horizon is actually greater at the North Pole than at the equator.[6] The greatest amount of snowfall that occurs on lowlands is perhaps on the shores of Lake Superior, where, according to Alexander Agassiz (as quoted by Sir C. Lyell), the average annual snowfall for fifteen years was seventy-two feet. Yet the snow never lay more than six feet thick on the ground, and disappeared completely in summer; although for four winter months the temperature was 5° F. below zero.[7] We may be quite sure, then, that beyond the mountain barrier of the Antarctic continent there are either lowlands and plateaux free from snow in summer, or an icy sea like that around the North Pole; but certainly no continuous ice-cap like that imagined by Mr. Croll.

It is in accordance with all these facts that the more remarkable indications of glaciation are always found in the vicinity of mountain ranges of sufficient altitude and extent to form condensers of the aqueous vapour and accumulators of snow, and that such indications point to a maximum of glaciation wherever the geographical position is such as to favour excessive rainfall. Thus in America the White Mountains of New England and the plateau north of the St. Lawrence have been the centres of the greatest glaciation on that continent; while in Europe, the Scandinavian mountains, and those of Ireland, Wales, and Scotland, together with the great central mass of the Alps, furnish the most remarkable evidences of extensive glaciation.

From these various considerations it is clear, that the increased cold of winter, when the eccentricity was great and the sun at its greatest distance during that season, would not of itself produce a glacial epoch, unless the amount of vapour to be condensed was also exceptionally great. Now the greatest quantity of snow falls in the Arctic regions in summer and autumn, and with us the greatest quantity of rain falls in the autumnal months. It is probable, therefore, that in all northern lands glaciation would commence at the time when autumn occurred in *aphelion*. All the rain which falls on our mountains at that season would then fall as snow, and, being further increased by the snow of winter, would form accumulations which the summer would not melt. As time went on and the *aphelion* occurred in winter, the perennial snow on the mountains would chill the summer vapours, so that they too would fall as snow, and thus increase the amount of deposition; and this would be still further increased from causes which have been admirably discussed by Mr. Croll.

The trade winds owe their existence to the great difference between the temperature of the equator and the poles, which causes a constant flow of air towards the equator. The strength of this flow depends on the difference of temperature, and this difference is now greatest between the South Pole and the equator, owing to the much greater accumulation of ice in the Antarctic regions. The consequence is, that the south-east trades are stronger than the north-east, the division between them being considerably on the north side of the equator. But just in proportion to the strength of the trade winds is the strength of the anti-trades--the upper current which carries the warm moisture-laden air of the tropics towards the poles. These are now strongest in the southern hemisphere, and thus they supply the moisture necessary to produce the enormous quantity of ice in the Antarctic seas.

But during the period we are discussing--of high eccentricity and the northern winter in *aphelion*--this state of things would be reversed. The South Pole, having its winter in *perihelion*, would be nearly free of ice, while the north-temperate zone would be largely snow-clad, and the north-east trades would therefore be far stronger than now. The south-westerly anti-trades would also be much stronger, and would bring with them the increased quantity of moisture, which is the chief thing required to produce a condition of glaciation.

The increased force of the north-east trades will have, however, another and totally distinct effect, which will still further increase the tendency to an accumulation of perpetual snow and ice. It is now generally admitted that we owe our mild climate and our comparative freedom from snow to the influence of the Gulf-stream, which equally ameliorates the climate of Scandinavia and Spitzbergen, as shown by the northward curvature of the isothermal lines, so that Trondhjem in N. lat. 63° 25' has the same mean temperature as Halifax (Nova Scotia) in N. lat. 45°. The quantity of heat brought into the North Atlantic by the Gulf-stream depends mainly on the superior strength of the south-east trades. When the north-east trades were the more powerful, its strength would certainly be much less, while it is possible, as Mr. Croll thinks, that almost the whole of it might be diverted southward, owing to the peculiar form of the east coast of South America, and would go to swell the

Brazilian current and ameliorate the climate of the southern hemisphere.

That some effects of this nature would follow from any increase of the arctic and decrease of the antarctic ice, may be considered certain; and Mr. Croll has clearly shown that in this case cause and effect act and react on each other in a remarkable way. The increase of snow and ice in the northern hemisphere is the cause of an increased supply of moisture being brought by the more powerful anti-trades, and this greater supply of moisture leads to an extension of the ice, which reacts in still further increasing the supply of moisture. The same increase of snow and ice diminishes the power of the Gulf-stream, and this diminution makes both summer and winter colder, and thus helps on still further the formation and preservation of the icy mantle. At the very same time these agencies are acting in the reverse way in the opposite hemisphere, diminishing the supply of moisture carried by the anti-trades, but increasing the temperature by means of southward ocean-currents.

We have now sufficiently answered the question, why the short hot summer would not melt the snow which accumulated during the long cold winter produced by high eccentricity and autumn or winter in *aphelion*, even though the amount of heat received from the sun in the whole year was exactly the same as now. The reasons are mainly two: first, that heat cannot be permanently accumulated, being continually carried away by winds and flowing water, while cold can be so accumulated, owing to the comparative immobility of snow and ice; and, secondly, because there are two great agencies, the winds and the ocean-currents, which are so affected by an increase of the snow and ice towards one pole and its diminution at the other, as to help on the process when it has once begun, and to produce by their action and reaction a maximum of effect, which without their aid would be altogether unattainable. To these we may add a third agency which we have not yet mentioned--that snow and ice reflect heat (as they do light) to a much greater degree than do land or water. The heat, therefore, of the short summer would be partly lost by reflection, and partly used up in melting or evaporating a certain portion of snow. Much of it too would be reflected from the upper surface of the clouds which, owing to intense evaporation at this season, would be very prevalent in the north temperate zone. Combining these various causes and effects, we have little difficulty in comprehending how the wonderful phenomena of glaciation in our own country, in the Alps of Central Europe, and over extensive regions in North America, were brought about.

Before quitting this part of our subject, we must briefly refer to the changes of level of the sea and land which occurred during the glacial period, and which Mr. Croll believes were mainly due to a shifting of the earth's centre of gravity owing to a transfer of the ice-cap from one pole to the other, thereby causing an accumulation of water around the heavier pole. There is evidence of many distinct periods of submergence. Mr. James Geikie believes that during the greatest extension of the ice, the land was somewhat lower than it is now; for he says that as the ice melted away in Scotland the sea occupied its place, following its retreating footsteps and laying down the boulder-clay; but this submergence did not exceed 200 or 260 feet. At a later period, the south of Scotland and Wales were submerged to a depth of about 1300 feet or more, up to which height marine shells are found in the drift. The land then rose again, with several pauses, till Britain became joined to the Continent, as it had been before the glacial epoch. Another submergence afterwards took place to a less amount, and then a slight elevation, bringing the land into the state in which it is now. A somewhat similar series of changes appear to have taken place in North America, but owing to the absence of marine shells the amount of subsidence is doubtful. Mr. James D. Dana believes that the land stood higher during the glacial period, and we have seen how important a factor this is in initiating glaciation and increasing its intensity.

On the assumption that an ice-cap, continually increasing in thickness, extended from lat. 50° in Europe and 40° in North America up to the pole, it has been calculated that a rise of the sea-level would be produced in the latitude of Edinburgh of about 800 or 1000 feet. But if, as we have endeavoured to prove, no such general ice-cap ever existed, but merely local ice-sheets collected over all high lands which were areas of great precipitation, then probably not more than one-fifth of this amount of ocean-rise would occur, sufficient to account for the submergence at the period of the formation of the boulder-clay, and for the various raised beaches at lower levels. The greater subsidences were probably local, and were perhaps due to the enormous weight of the accumulations of ice over given areas. Owing to the earth's crust giving way slowly to such strains, subsidence would only begin when the ice-sheet had nearly attained its maximum extent, and would probably continue for some time while it was diminishing; and this seems to accord very well with the ascertained facts.

It will now be seen that the theory, by which the glacial epoch is accounted for, explains also those curious indications of intercalated warm periods to which we have already referred. For after a lapse of about 10,500 years the *aphelion*, or time of minimum sun-heat in the northern hemisphere, would have gradually changed from winter to summer, and the *perihelion*, or time of maximum sun-heat, from summer to winter. The result would be, that our winters would then be much warmer and our summers somewhat cooler than they are now, and we should thus have a kind of perpetual spring. At the same time the south-westerly winds would be less strong, and would bring less vapour, while the Gulf-stream would be much stronger, and this would still further ameliorate our climate and help to get rid of all the snow and ice, till at the height of this mild epoch both would be probably unknown. We find then that our theory necessitates the existence of these warm interglacial periods, the evidences for which were once such a puzzle to geologists: and we cannot but admit that this circumstance is an additional support to the theory. We have now, however, to consider a series of facts of the most interesting nature, which bring out this aspect of the value of our theory in a still more convincing manner.

Evidence of the former occurrence of Warm Climates in Arctic Regions.--That a milder climate prevailed in the Arctic regions, at a not very remote epoch, is proved by two distinct kinds of evidence. We have, first, the remains of large herbivorous mammalia, such as the mammoth, woolly rhinoceros, bison, and horse, found in the icy alluvial deposits of Northern Siberia, and sometimes preserved with the skin and carcass entire. These remains occur over the whole area of Northern Asia, and are abundant in the Liakhov Islands, nearly ten degrees north of the Arctic circle; and though the animals may have been overwhelmed by floods and buried in icy deposits, it is generally admitted that, in order to supply such large beasts with suitable food, the climate of Northern Siberia must have been less severe than it is now.

Less doubtful proofs of a mild climate are however afforded by the numerous discoveries of unfossilized trees in the arctic lands of North America. Remains of large pine-trees were found abundantly by Sir Robert M'Clure in a ravine on the west coast of Baring Island (about 73° N. lat.). A piece of one of these trees which was brought home was twelve inches in diameter, and was estimated to have seventy rings of annual growth. Similar remains have been found in Prince Patrick Island, and in Wellington Channel, much further north; and all the observers agree that these trees grew where they are found, although they are more than 700 miles farther north than similar trees grow now. Trees capable of being still used as fuel are found in about the same latitude on the Liakhov Islands off the north coast of Siberia. Here we have undoubted proofs of a considerably milder climate in the extreme north; and the perfect condition of preservation in which the trees are found renders it probable that they are to be referred to one of the last mild phases of the glacial epoch.

Very much more remarkable, however, are the remains of a luxuriant flora of the Miocene period, which have been found in great abundance in many arctic localities, especially in Iceland, Greenland, and Spitzbergen, with smaller deposits at the Mackenzie River, in Banks's Land, and in Grinnell Land, lat. 81° 45' N. This flora consisted of conifers in great abundance, more than fifty species being known, about two hundred and thirty flowering plants (Monocotyledons and Dicotyledons), and seventeen species of ferns and horsetails. Taking the plants found in Greenland, in 70° N. latitude, as well representing the general character of this flora, we find seven species of oaks, two planes, three beeches, four poplars, two willows, two vines, a walnut, a chestnut, a sassafras, a liquidambar, a magnolia, a plum, and several shrubby plants, together with sequoias allied to the mammoth trees of California, and salisburias allied to the curious gingko-tree of Japan--in all one hundred and thirty-seven species of trees and shrubs, often beautifully preserved, and many of them possessing finely developed foliage. Of course we cannot suppose that we have here collected in one spot the entire flora of the country, as in the herbarium of a botanist, and, looking upon it merely as an accidentally preserved fragment, we are led to conclude that the flora of Greenland was probably richer then than that of any part of Europe is now, and perhaps as rich as that of the Northern United States or Japan.

Much further north, in Spitzbergen, in 78° and 79° of north latitude, and now one of the most inhospitable climates on the globe, a fossil flora equally rich in species has been found, but of a somewhat more northern character. Besides pines, sequoias, and the swamp-cypress of the Southern United States, there are oaks, poplars, planes, limes, a hazel, a birch, and a walnut, and also water-lilies, pond-weeds, and an iris. Several of the species are identical with those of Greenland.

In Grinnell Land, within eight degrees of the pole, the plants found consist of ten conifers, including the swamp-cypress, with a poplar, a birch, two hazels, an elm, a dogwood, a water-lily, a reed-grass, a sedge, and a horsetail, indicating probably a climate not very unlike that which now prevails in Scotland.

In the other localities referred to, as well as in Alaska and the Liakhov Islands, less extensive remains of a similar flora have been found. Another occurs near Dantzic in 55° N. latitude, containing the swamp-cypress, sequoias, oaks, and poplars; along with a few more southern forms, as laurels, figs, and cinnamons. In the Isle of Mull, in Scotland, a fragmentary deposit of the same age has been preserved, with a hazel, a plane, and a sequoia, and similar plants occur at Bovey Tracey in Devonshire. It is however in Central Europe, at Œninghen in Switzerland, and near Breslau, north of the Carpathians, that the richest deposits of Miocene plants have been found, and some of these are identical with Greenland species, while many others are closely allied. There is in fact the clearest evidence that one characteristic flora then covered the whole of the north temperate and Arctic zones, with just about as much difference as now exists between the vegetation of France and Norway or the Southern United States and Canada. This flora bears more resemblance to that now inhabiting Eastern North America and Eastern Asia, than to the flora of Europe; and though it clearly shows differences of climate according to latitude, these differences are far less than such as now exist, indicating a much greater uniformity of temperature.[8]

We have already explained that, if glacial epochs were mainly due to great eccentricity of the earth's orbit, they must always have alternated in each hemisphere with correspondingly mild periods, each lasting about 10,500 years; and we have shown that evidence of such intercalated mild periods is actually found, both amid the records of the last glacial epoch in Scotland, Switzerland, and North America, and in the still existing remains of large unfossilized fir-trees in the highest Arctic regions. This glacial epoch has been assumed to have been caused by the first period of high eccentricity which we meet with in our backward survey-- that which occurred 200,000 years ago. Still further back, we meet with no equally high eccentricity till 850,000 years ago, when it considerably exceeded that of the last glacial epoch, amounting to near 6 3/4 instead of 5 millions of miles, and causing a difference of about thirty-six days between the length of winter and summer. At this time there must have been a correspondingly severe glacial epoch, with alternating periods of very mild spring-like climates, and it is to one or other of these alternating mild periods that we impute the various Miocene floras that have been discovered in so many places in the Arctic regions and in the North temperate zone.

As we have already sufficiently explained how the mild climate, approaching to a perpetual spring, was caused in the northern hemisphere at the time when the southern hemisphere was highly glaciated, we need only say here that the greater degree of eccentricity would produce at least a correspondingly greater effect. But we know that considerable changes of land and sea have occurred since the Miocene period, and we may well suppose that these were sometimes favourable to the climate of the Arctic regions. Such would be the case if the extent of land in the Antarctic regions were greater than now, admitting of a much greater extension of the snow and ice. This would increase the flow of the Gulf-stream, and of other currents bringing warm water to the north to an enormous extent,--and if we also suppose that there was less high land near the north pole, and some difference in the American coast line, admitting the warm currents more freely into the Arctic regions; and if we further remember that during winter the whole north temperate zone would be kept warm by the nearness of the sun at that season,--it is not difficult to understand that the cold of the Arctic winter might be so ameliorated that, notwithstanding the four months' absence of the sun, the winter's cold of Spitzbergen might never exceed that of the north of Scotland at the present time.

It is now generally admitted by botanists, that the three or four months' polar night would be no serious obstacle to the growth of a luxuriant forest vegetation, if the temperature and general climate were otherwise suitable. The hothouses in the Botanical Gardens of St. Petersburg are thickly matted up for six months, yet the prolonged darkness does not prevent even tropical palms from being successfully cultivated; while in Arctic lands many shrubby plants are covered up with snow for more than that period. A far more important element in the growth of broad-leaved dicotyledonous trees is a tolerably calm atmosphere. Long-continued cold winds and gales are highly prejudicial, and it is to these that the absence or scarcity of trees in many of the stormy islands of the southern hemisphere is probably due. It is this too that dwarfs the Arctic trees; and it is probably their power of resisting long-continued winds, that enables tall coniferæ to extend further north than equally lofty deciduous trees. Now it is an interesting fact, which appears to have escaped the attention even of Mr. Croll, that during the mild periods caused by high eccentricity there would be a great diminution of wind, owing to the much greater uniformity of temperature over wide areas and to the weakness of the trades and anti-trades of the northern hemisphere. That there must have been such an absence of violent winds during the summer, is evident to any one who looks at the foliage of the Arctic trees figured in Heer's work, specimens of which may be seen in the geological gallery of the British Museum. The leaves of the oaks, planes, chestnuts, vines, and poplars, are larger than those of most European species, and plainly tell of a comparatively tranquil atmosphere; and it is certainly a most suggestive fact, that the same theory, which accounts for so many of the past changes of our earth, explains also this undoubted feature of the Arctic climate in the Miocene warm period.

The difficulty of associating such luxuriant vegetation with the long polar night and its accompanying severe cold, has led many writers to maintain the necessity of a considerable decrease in the obliquity of the ecliptic, while others have supposed that the position of the pole itself may have changed. But astronomers and physicists both deny that such changes have occurred to anything like the requisite amount, while the latter change would not produce the desired effect, since there are proofs of a nearly simultaneous mild climate all round the polar area.[2] The changes of eccentricity, and of the phases of perihelion and aphelion, are, on the other hand, universally admitted and calculable phenomena; while their effects on the accumulation of snow and ice, on the winds and on the ocean currents, have been worked out in so masterly a manner by Mr. Croll, that they are rapidly taking their place among the established deductions of physical science. If we add to these such moderate changes in the distribution of sea and land, the outline of coasts, and the elevation of mountains and plateaux, as we know to have occurred again and again during the Tertiary period, we obtain a combination of causes which seem fully adequate to have brought about those wonderful changes of climate, manifested on the one hand by the recent glaciation of our own islands, and on the other by the luxuriant Arctic vegetation of the Miocene age.

Indications of Glacial and Mild Climates throughout Geological Time.--It will of course occur to the reader that, if the extreme eccentricity 850,000 years ago was the cause of the mild climate in the Arctic regions, and its accompanying luxuriant flora, there must have been also, each alternate 10,500 years, a glacial epoch of extreme intensity; and it may be asked, where are the proofs of such periods of glaciation? The answer is that there are some indications of such a glacial epoch, and, though very scanty, they are such as we can alone expect to find. The evidence of the last glacial epoch, which is more especially convincing, is that of the superficial effects produced by the ice--the striations, the *roches moutonnées*, the moraines, the travelled blocks, and the 'till.' But these have only been preserved to us here and there, because their formation is so recent, and because they once covered the whole country. The greater part of them have been destroyed; and that any traces of them still remain is probably due to the fact that, since the last glacial epoch passed away, there has been a period of very low eccentricity, and consequently great stability of climates, and comparatively little denudation. No fragment of any such evidence from the remote Miocene glacial epoch could possibly reach us, because there has been, in almost all the time intervening between it and our last glacial epoch, an amount of eccentricity always much greater than now, and on the average nearly double. The whole lapse of time has therefore been a continued series of alternate periods of mild and severe climates, often culminating in lesser glacial epochs and necessarily leading to a great denudation and an almost complete remodelling of the earth's surface. There are only two kinds of evidence of these remote glacial epochs that can possibly remain to us,--beds containing glacial fossils, and travelled angular blocks imbedded in marine or lacustrine deposits. The former have not yet been recognized, though they probably exist, and it is quite possible that some deposits of Arctic shells, classed as glacial, may really belong to this period. The shells of warm seas would undoubtedly suffer modification and extinction by change of climate, but those of cold regions might perhaps undergo little change; for though the polar seas of mild epochs might never be frozen, neither would they ever rise in temperature much above the freezing-point, except in comparatively shallow water. Many of the Arctic mollusca live at considerable depths and have a wide range, and it is not improbable that the Arctic shell-fauna has continued almost unchanged from the Miocene period. It is therefore quite possible that some of the numerous deposits with shells, which have been found in the Arctic regions at elevations of from 100 to 500 feet above the sea, may belong to the Miocene period. The same difficulty occurs in the interpretation of the crag and associated deposits of our own country. The presence of a large proportion of living and Arctic shells has been held to prove a more recent origin, whereas it may indicate an intercalated cold period at an earlier date. Geologists almost always form their conclusions as to the age of these fragmentary deposits on the assumption that there was a continuous deterioration of climate from the Miocene through the Pliocene to the Glacial epoch. If, however, as we believe, there have *always* been alternations of warm and cold periods, a very different interpretation may be placed on the facts, and some difficulties may be overcome.

The other kind of evidence, that of travelled blocks, is however found in the Miocene deposits of Central and Southern Europe. In Northern Italy, near Turin, there is a sandstone formation full of Miocene shells, intercalated among whose strata are beds of conglomerate, containing huge angular blocks of serpentine and greenstone, sometimes more than twenty feet long. Some of them are partially striated and polished, and similar rocks occur in the Alps about twenty miles off.

When we go back to the Eocene period, we find

indications of a decidedly more tropical climate than the Miocene, in fact quite as warm as that of any earlier period, so far as we can judge by organic remains, for palms, turtles, large snakes, and crocodiles, then inhabited England. Yet in this period we also have indications of ice-action, in an extensive deposit of finely stratified sandstone several thousand feet in thickness, extending from Switzerland to Vienna, and quite destitute of organic remains, but which contains in several places enormous blocks, either angular or partly rounded, and composed of oolitic limestone or granite. Near the Lake of Thun one of the granite blocks of this formation is of huge dimensions, being 105 feet long, 90 feet wide, and 45 feet thick. The granite is red, and of a peculiar kind, which cannot now be matched anywhere in the Alps. Similar erratics have been found in beds of the same age in the Carpathians and in the Apennines, indicating probably an extensive inland European sea, into which glaciers descended from the surrounding mountains.

Going back from the Eocene all through the Secondary formations, the organic remains are such as to indicate warm European seas, and there are no satisfactory proofs of ice-action. But when we reach the remote Palæozoic formations, we lose all clear evidence of very high temperatures, since even the wonderful coal-flora is now generally admitted to be indicative of a mild or warm uniform climate, but by no means necessarily of a tropical one. Here again we meet with unmistakable signs of ice-action in the Lower Permian conglomerates of the west of England. These contain partially-rounded or angular fragments of various rocks, with striated or polished surfaces just like the stones of the 'till.' These blocks lie confusedly bedded in a red unstratified marl, and can often be traced to Welsh rocks, from twenty to fifty miles distant. This remarkable deposit was first adduced by Professor Ramsay as indicating a remote glacial epoch, and, after a personal examination of it on the ground, Sir Charles Lyell agrees that this is the only possible explanation that, with our present knowledge, we can give of it.

Numerous examples of erratic blocks, which seem to indicate ice-action, occur in the Carboniferous formation in Scotland, in France, in Nova Scotia, in Ohio, in India, and in Australia. The Old Red Sandstone and Silurian formations contain similar blocks and boulders; while Principal J. W. Dawson, of Montreal, who is not an extreme glacialist, and whose opinion is therefore unbiassed, believes that we have evidence of ice-action in temperate latitudes as far back as the Huronian age--that is beyond our Cambrian formation.

The only evidence wanting to complete the proof of glacial epochs having occurred repeatedly throughout all geological time, is the discovery of deposits of arctic marine shells similar to those of the drift, which have sometimes been raised more than a thousand feet above the sea. But here again it is a question whether such deposits can be recognized if they exist. The alternate periods of about ten thousand years of mild, and of glacial conditions, are so short geologically, that the marine deposits formed during a series of such changes may be represented by alternate bands or strata in one deposit, and the fossils of both periods may be more or less mingled together. Even if the deposit formed during a phase of glacial conditions is sufficiently distinct in composition to be separated from adjacent beds, its comparative poverty in organic remains, or their small size, will not be imputed to cold, because geologists have not yet recognized the constant alternation of short periods of mild and glacial climates as an established fact, but are accustomed to consider a glacial epoch to be one long-continued period of generally glacial conditions. Again, it must be remembered, that during the cold periods denudation will be greatly checked, so far as the carrying of the denuded matter to the sea by running water is concerned; while during the succeeding warm period it will be at a maximum, the melting of snow and ice being added to the normal rainfall, while there will be abundance of loose materials ground off by the ice, to load the rivers with great volumes of sediment. It follows, that marine deposits representing periods of glaciation will always be very scanty, as compared with those of the succeeding warm periods, and this is another reason why any deposits of this kind which do exist may easily have been overlooked.

But if the evidence of remote glacial epochs is rare and

fragmentary, that of the occurrence of warm periods in the Arctic regions is frequent and ample. Besides the wonderful Miocene flora already described, there is a somewhat older one of the Upper Cretaceous age in Greenland, containing besides abundance of dicotyledons--such as figs, magnolias, myrtles, and the sub-tropical genus Myrsine-cycads, conifers, and numerous ferns, one of which is a tree-fern, with a thick stem which has been found in the Greensand of England.

In the same locality in Greenland (70° 33' N. latitude, and 52° W. longitude), and also in Spitzbergen, a more ancient flora of the Lower Cretaceous age has been found, differing widely from the preceding in the great abundance of ferns, cycads, and conifers, and the scarcity of dicotyledons, which are represented by a single species of poplar. Among the ferns many belong to the genus Gleichenia, now entirely tropical.

Proofs of a mild Arctic climate in Jurassic times are found in the rich flora of this age in East Siberia and Amur-land, with less productive deposits in Spitzbergen, and at Ando in Norway within the Arctic circle. But more remarkable are the ammonites and the vertebræ of Ichthyosaurus and Teleosaurus found in the Jurassic beds of the Parry Islands in 77° N. latitude.

In the still earlier Triassic deposits of Spitzbergen, species of nautili and ammonites also occur.

True coal of the Carboniferous period has been found

at Spitzbergen and at Bear Island, on the north coast of East Siberia, and it contains Calamites and Lepidodendrons, with large spreading ferns. Marine deposits of the same age contain large stony corals; while the more ancient Silurian limestones, which are widely spread in the high Arctic regions, contain abundance of corals and shells of Cephalopodous mollusca, like those of the same formation in the temperate zones.

It must undoubtedly be admitted, that this connected series of records of the animal and vegetable productions of the Arctic regions, extending over the whole vast period from the Silurian to the Miocene, inclusive, and widely scattered over the Arctic and sub-arctic zones, indicates rather a constant mild climate than an alternation of warm and cold periods; and it is therefore not surprising that this view should be generally adopted. We must remember, however, that there are yet vast gaps in the record, representing long ages of which we know nothing; while the condition of the Arctic regions during cold periods would certainly be such as not to favour the formation of stratified deposits, or the preservation of animal or vegetable remains. Denudation by water would be almost wholly checked, except perhaps where great northward-flowing rivers brought down the products of warmer lands; while the masses of debris carried by glaciers and icebergs, even if preserved to our times, would rarely contain fossils.

In view of the important part played by extensive tracts of high land in producing glaciation, it is worthy of consideration whether, in the absence of such conditions, anything like a severe Arctic climate could exist, even during periods of high eccentricity with winter in aphelion. Mr. Croll has well remarked, that the influence of the lofty ice-clad mass of Greenland on the climate of the northern hemisphere is overwhelmingly great, and that were it wholly removed the resulting amelioration of climate, even now, would be almost magical. Some portion of the milder climate in the Arctic regions may then be due to the absence of any high land in districts of great precipitation; and we may easily imagine such an arrangement of the land as to concentrate all the heating power of the oceanic currents in a limited area, and thus to produce a mild or even warm climate on one side of the pole, while the other side experienced an Arctic winter. And if at this time high land prevailed, as now, round the south pole, the powerful influences of winds and currents, already described, would be kept constantly at work, so as to intensify the glaciation of the Antarctic, while ameliorating the climate of the Arctic regions. Such favourable geographical conditions might possibly keep up a warm arctic climate during an entire phase of high eccentricity, and if these conditions prevailed in several geological periods, the abundant indications of luxuriant floras and sub-tropical faunas in high northern latitudes would be to a great extent explained.

Let us now briefly summarize the facts of this strange history. The geological record is best known to us in temperate latitudes, and the series of extinct animals and plants it has brought to light indicates almost always warmer climates than at present. The Pliocene was a very little warmer, the Miocene considerably warmer, the Eocene almost or quite tropical. Further back we have no proofs of any increasing warmth; and it is generally admitted that even the carboniferous flora does not imply a climate at all warmer than the Eocene. But at a comparatively very recent period, just at the close of the Pliocene, we have irresistible proofs of intense cold in the northern hemisphere, which reduced the northern half of our own islands, and much of Europe and North America, to the condition in which Greenland is now. But there are also indications that this arctic climate alternated with milder intervals; and further, that in Miocene, Eocene, and many older periods, distinct glacial epochs occurred, which may have been as severe as that we have recently gone through. Then we have another series of still more startling climatic changes, in the warm climates of the Arctic regions. These have been proved to occur, first, probably, at or near the time of the glacial period; then in Miocene, Upper Cretaceous, Lower Cretaceous, Jurassic, Triassic, Carboniferous, and Silurian times. There is also evidence that some similar changes occurred in the southern hemisphere, of which, however, our limited space has not permitted us to give any account.

Now the whole series of these wonderful changes can be explained by a full consideration of the influence of certain astronomical facts--the eccentricity of the earth's orbit, and the precession of the equinoxes--combined with the physical properties of snow and ice in storing up cold, and the action and reaction of these on the winds and the ocean currents; the whole being further modified and intensified by changes in the distribution, and especially in the elevation, of the land in temperate and polar regions.

The objection to the theory is, that it accounts for too much. It not only explains all the changes of climate of which we have evidence, but it necessitates a whole series of changes of which we have no direct evidence. Some attempt has been here made to explain why the record of such changes is unlikely to have been preserved, and why, in cases where it has been preserved, it may nevertheless have been overlooked. The imperfection of all our records of the past is too well known to geologists, for this difficulty to have much weight with them; but we may further point out that none of the alternative hypotheses yet suggested at all remove this difficulty. If the pole had shifted its place any number of times to bring Greenland or Spitzbergen into warm latitudes, the Arctic regions must still have been *somewhere*, and the difficulty is, that no Arctic remains are *anywhere* found beyond recent times. And if we postulate any amount of change in the obliquity of the ecliptic (as advocated by the late Mr. Belt), we still have to trust to differences of eccentricity, and of winter or summer in perihelion, to produce glacial epochs and warm arctic climates alternately, and this leaves the problem exactly where it is now. As to the theory of a cooling earth, even if it were not totally inadmissible on physical grounds, it would leave the glacial epoch itself--the great starting-point in the complex problem of terrestrial climates--totally unaccounted for. We claim, therefore, that the known facts of 'eccentricity,' when properly applied, do serve to explain the known changes of our climate in past time; and that this is really the only working hypothesis now available, since all others have to make assumptions which either astronomers, physicists, or geologists will not grant. It were much to be wished that palæontologists would keep this theory in view when studying in detail the subdivisions of any formation, with the object of ascertaining whether such evidence of changes of climate as it requires may not sometimes have been overlooked.

Notes Appearing in the Original Work

[1]'Antiquity of Man.' 4th ed., pp. 340-348.[2]'The Great Ice Age,' p. 177. [3]Heer's 'Primæval World of Switzerland,' vol. ii. p. 296. [4]'Proc. Royal Geog. Soc.' vol. xxi. p. 276.[5]'On the Glacial Epoch,' 'Geological Magazine,'

July, August, 1864. [6]'Philosophical Magazine,' February, 1870. [7]'Principles of Geology.' Eleventh edition, i. p. 290. [8]Some writers believe that the Arctic floras were not contemporaneous with those of Temperate Europe, the former being Eocene and Lower Miocene, while the latter are of Lower and Upper Miocene age; but this point is of no importance for our present object, which merely is, to show the occurrence in former ages of a vegetation in very high latitudes characteristic of warm regions. [9]Rev. Samuel Haughton's 'Notes on Physical Geology.' Read before the Royal Society, and published in 'Nature,' vol. xviii. p. 266.